Bibliografische Information der Deutschen Nationalbibliothek:

Die Deutsche Bibliothek verzeichnet diese Publikation in der Deutschen Nationalbibliografie; detaillierte bibliografische Daten sind im Internet über http://dnb.d-nb.de/ abrufbar.

Impressum:

Druck und Bindung: Books on Demand GmbH, Norderstedt Germany
ISBN: 9783668763852

Johann Gruber-Schmidt

Biogas Conversion to Dibutyl ether and Dimethyl ether supporting ETS

GRIN Verlag

1 Biogas, Biomass Gas, Carbon Dioxide Conversion to Dimethyl Ether and Dibutyl Ether based on Emission Trade System

Johann Gruber-Schmidt

Abstract: In this short report we describe the biogas conversion to dimethyl ether. The generation of biogas in a structured biogas plant based on biomass wet biomass. The simple model of a bioreactor helps to build up a simple dynamical model for predicting the biogas generation. In this chapter the gas components of biogas will be described including pollution and additional components beside methane and carbon dioxide. After collecting the biogas gas, the cleaning, the preparation and the conversion to dimethyl ether is the next step. The conversion of biogas to dimethyl ether (DME) is also known as liquefaction of biogas gas, if we use biomass biomass often called biomass to liquid fuel. Additional the necessary and sustainable property of Zero Emission is described and analyzed. The main influence on the ETS (= emission trade system) is analyzed and shown.

Keywords: Biogas, Gas Cleaning, Gas Preparation, Methanol Synthesis, Dimethyl Ether, Dibutyl Ether

1.1 Content

1.2 Biogas Plant (Structure, plant and properties)

Biogas Plants and Biogas technology today is well-known and standard. The idea behind biogas plants is to build up a large fermentation reactor system similar to the stomach of a cow. [25]

Whether plastics no inert parts can be used for conversion to biogas gas. The composition of biogas mainly consists of methane and carbon dioxide. The generation of biogas is connected with the existence of water, microbial bacteria and nutrients. The water collected in the biogas plant and returned to the biogas plant is called leachate. This makes sense, because so the nutrients and microbes can be saved.

Modern biogas plants consist of storage systems for the substrate, substrate preparation, where the leachate is collected and the leachate cannot drain into the ground water. The leachate is supported by a leachate piping network into the digester.
The digester, where the fermentation and conversion to biogas takes place, and a storage for the digestate. [25]

Every section of the biogas plant is divided into structures connected to the gas collection pipes. The gas will be sucked out of the digester. The gas pipes are lead to collectors and the collectors are connected with pipes leading the biogas out of the biogas plant. This marks the limit of the biogas plant. In most cases biogas is burned in gas engine to generate electric power and thermal heat, well-known as combined heat process (CHP).

Understanding the biogas plant as a bioreactor biogenic biomass in combination with water and microbes are needed. Supporting the model of a bioreactor leads to the demand of preparing separating and recycling parts like plastic, metals or sand. In the end the aim should be to have inert material lower than 5%. Under these conditions the biogas plant is nearly a perfect stationary bioreactor with a high efficiency generation over a long-time range biogas gas. [24]

The biogas plant consists of digester closed with a double flexible membrane to collect the biogas and to provide a high density of the digester to the environment.

Biogas plants from former times are acting as a black box. Measuring the biogas volume flow and the gas composition, and assuming the volume of the biogas plant enables assuming the possible biogas production. Because of the structured design and the definite types of biomass and biogenic biomass composition there is the possibility to control all parts of the biogas plant, to optimize the operation and to operate on a high safety level.

A modern and new biogas plant designed as a bioreactor is very simple in erection, easy in handling and operation. It includes the storage of the substrate, the preparing of the substrate and the microbial fermentation process converting biogenic substrate to biogas gas. The biogas plant is well structured and supported with a piping network.

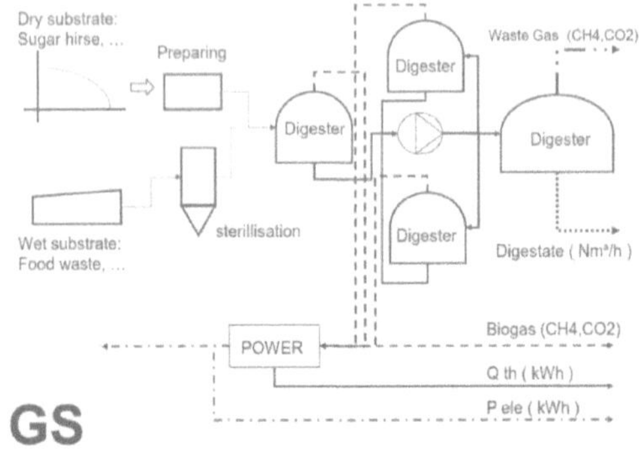

Figure 1: Simple structure of a Biogas Plant operating on the basis of wet fermentation as shown in the project Hagenbrunn [2]

According to the observation and measurements the composition of the substrate used in the biogas plant is given by: (first classification)

Storage for as long time:
Biogenic Parts 70% - 80%
Water 20% - 30%

Samples: maize straw, sugar sorghum straw, corn straw, horse manure
Structure: concrete open silo [24]

Storage for a short time:
Biogenic Parts 20% - 30%
Water 70% - 80%

Samples: vegetables, biomass food, potatoes, onions, ...
Structure: Tanks, pumpable [24]

The better the structure of the biogas plant is done, the more we know about the biogenic mass and mass distribution and substrate type used in the digester, the better the gas generation can be controlled and optimized. Under this condition we can reduce the risks for the environment.

Under recycling and separating the municipal biomass into recycling streams and biogenic biomass with a very low content on inert material, the biogas plant is acting like a stationary bioreactor. Inert material can be separated in different material streams again. One advantage is the reduction of the biogas plant volume, the other advantage is the reduction of hazard biomass in the biogas plant and on the economical side the operating costs of an advanced biogas plant are very low. The advanced biogas plant in combination with recycling and separation leads to an optimized gas generation.

1.3 Biogas (Phases of Gas Production)

The typical phases of gas production inside the biogas plant have been studied over long years and the typical phases can be described for the continuous reactor:

I Phase	Hydrolysis
II Phase	Fermentation
III Phase	Acetogenic Phase
IV Phase	Methanogenic Phase
V Phase	Storage of digestate

The duration of the different phases can be measured and qualitative empirical described in the time measure "day": (estimation, depends on the type of substrate and the preparation):

I Phase	1 day
II Phase	7 days
III Phase	10 days
IV Phase	30 days (end of the biogas production)
V Phase	180 days (resting of digestate)

The gas production of the phase II and IV are the highly productive phases and are in main interests.

1.4 Gas Components Biogas

The gas composition of biogas and the volume flow depends on the different phase of the biogas plant. The design and the calculation for the conversion of biogas can be taken from the phase III and IV. Important is the gas mixture of methane and carbon dioxide, mainly methane.

Methane (CH_4)	40% up to 70%
Carbon Dioxide (CO_2)	30% up to 60%

Additional additive components produced during the process of biogas gas, which are also sucked out (additives are higher than pollution):

Non-methanogenic organic components		0% up to 0,25%
VOC		0% up to 0,1%
CO		0% up to 0,2%
H_2S		800 ppm
H_2O	100% saturated	0% up to 14% (opera.Temp.)
N_2		0% up to 3%

Pollution observed and measured in biogas depend on the substrates used:

Benzene	30 ppm
Toluene	300 ppm
Organosulfur	50 ppm
Siloxane	300 ppm

Inside the biogas side the operation temperature is varying in the range of 37°C up to 57°C. The lower temperature enables the growing of the microbial bacteria, the higher temperature supports the chemical conversion reactions. The biogas is taken out of the biogas plant with a vacuum compressor, inside the biogas plant a small vacuum pressure (p ~ 25 mbar) is given, after the vacuum compressor a higher pressure (p ~ 1.5 up to 2 bar) is given. Cooling down the biogas gas, the steam changes the phase and is removed as condensate. Therefor a condensate trap has to be used. The gas components methane and carbon dioxide define the lower caloric heat value, the additives and pollution define the cleaning and preparation of biogas gas.

1.5 Gas – Measurement

The gas measurements on the biogas plant are necessary and needed to control the conversion to biogas to optimize the efficiency and the production of biogas gas. The measurements include the concentration of methane, carbon dioxide, and nitrogen. Additional the organic sulfur compounds, the chlorine compounds and fluorine compounds and the halogen organic compounds. The measurement is done continuously.

Beside the main parts of the biogas the pollution is important for the emission to the environment a continuous measurement is needed: sulfur compounds, fluorine hydrogen compounds (HF), chlorine hydrogen compounds (HCL), siloxane compounds (X-Si-X), volatile organic carbon compounds (VOC).

The pollution in the biogas observed and measured is given by benzene, toluene, siloxane, sulfur hydrogen, and odor compounds. The pollution is important for the conversion of biogas to electricity and heat and fuel. The quality of the fuel is an important property and has to be reached during the conversion process and it has a deep influence on the costs and the plant components. These measurements are not needed continuously – it is needed discontinuous.

Converting biogas to heat, it will be burned in a combustion chamber, measurements of fluorine, chlorine, sulfur, carbon monoxide, nitrogen oxide are done continuous. Additional measurements have to be done: dioxin, furan, polycyclic aromatic hydrocarbons (PAK), polycyclic halogen aromatic hydrocarbons (HPAK). To reduce the emission to the environment, we have to use exhaust gas cleaning and preparation, in some cases high temperature combustion is needed.

Converting biogas to the liquid fuel dimethyl ether demands a definite purity of the biogas according the ISO 16 618:2013, ASTM 7509-14A und ASTM 7509-14B defines the allowed impurities and pollution in the biogas. [31]

1.6 Chemical Reactions in the Dynamic System

To understand the production off biogas the main components are listed up in simple chemical reaction equations:

$$C_6O_5H_{10} + H_2O \longrightarrow 3CO_2 + 3CH_4 + Q(-\ 205 kJ/mol) \qquad [E\ 1]$$

$$C(org) + H_2O \longrightarrow \frac{1}{2}CO_2 + \frac{1}{2}CH_4 + Q(+\ 9.5 kJ/mol)$$

For the biogas the following components are important: water, organic compounds (cellulose) and carbon. The chemical reactions show the production of carbon dioxide and methane. These simple equations give a first inside look about the processes taking place in the biogas plant. The main problem for every biogas plant is the knowledge of the distribution on the organic and biogenic biomass and the water distribution (we often speak from load in the digester: 2 up to 4 g/l). Water is the main important compound, therefore in all model the solved carbon in water the so called aqueous carbon is needed. This carbon is the main compound for the generation of methane and carbon dioxide. [24]

According to the measurements and observations some useful design values and construction values can be listed up:

400L / CH_4	/ kg	dry biomass
180 L/CH_4	/ kg	wet biomass
200 up to 280 L/CH_4	/kg	wet biomass

The caloric heat value (lower) (LHV [kJ/kg; kJ/Nm ^{3}; kWh/kg; kWh/Nm^3]) of different fuels compared with each other from [20]:

Coal	22 MJ/kg
Oil	40 MJ/kg
CNG	11 kWh / Nm ^{3}
Lean gas:	4 – 6 kWh/Nm^3
Biogas	4 – 6 kWh/Nm^3
DME	8,3 kWh/kg
DEE	9,8 kWh/kg
LPG	11,2 kWh/kg
Diesel	10,0 kWh/kg

Biogas is also a lean gas. A lean gas is a gas with a very low caloric heat value in the range of 1,5 kWh/Nm ^{3} up to 6 kWh/Nm ^{3}. It is well known to burn lean gas in a combustion chamber. Another possibility is the conversion to liquid fuel. This development is supported by low costs in the operation of a biogas plant, and low costs for the biogas itself. The costs of the produced biogas are the sum of costs for opening a biogas plant, the leachate piping grid, the gas suction piping grid, the safety components of the plant, the preparing of the recycled leachate. Additional the operating costs like personal, maintenance is added. It makes sense to separate the biomass and to structure the biogas plant, so that the optimum gas generation can be achieved for a long-term rate. The costs of the generated biogas influence the costs for the liquid fuel. At this point we have to mention the Zero Emission property again increasing the efficiency and sustainability by using the generated carbon dioxide as a raw material for the liquid fuel.

1.6.1 Biogas Plant – Dynamic Model

The Biogas model is a very simple but effective model, with a high empiric part. The model is based on a system of differential equations and predicts the methane generation:

The model is a result of practical experience. The idea is to make an assumption of the DDOC (decomposable organic carbon) based on the measurements, the observations of existing biogas plants. At least the measurements of the concentration of the biogas and the volume flow of the last operation year have to be taken into account. Under this condition two data sets have to be combined the actual operating year and the operation year ago. With this simple model the unknown structure of biomass distribution and biomass types can be neglected. The result helps to gain first hints for the production of biogas and the running time of the biogas plant.

Beside this simple model there have been developed a lot of models but all models suffering on the lack of data and information of the biogas plant and structure lead to results with high uncertainty.

The result is: if the biomass is prepared and separated and the structure of the biogas plant is designed according to the biomass type, the biogas plant will act like a large stationary bioreactor with large volume, a small surface and the gas generation can be optimized. Influencing the water content, the microbial distribution, the nutrient and the biogenic biomass type, the gas generation will lead to high volume flow with high methane concentration.

1.6.2 Dynamic Model of the Bioreactor [22]

A dynamic model for the calculation of the generation of biogas is based on the continuity equation, the energy equation, and the growing equation, well known in the classical model of Monod. Substrate S, Product X and the saturation constant K, decay rate μ:

$$\frac{dX}{dt} = \frac{mS}{K+S}X \qquad \text{[E 2]}$$

As described before the solid carbon in biomass, organic biomass and biogenic biomass is solved as aqueous carbon in water. This liquidated carbon is important for the generation of methane and carbon dioxide: continuity on the solid carbon mass, and the chemical reaction equation for the aqueous carbon are given:

$$\frac{dC_s}{dt} = - K_s C_s \qquad C_s + H_2O \longrightarrow C_{aqu} \qquad \text{[E 3]}$$

Continuity:
Solid organic carbon:

$$\frac{dC_s}{dt} = - /_n K_s C_s \qquad \text{[E 4]}$$

Continuity:
Liquid organic carbon:

$$\frac{dC_{aq}}{dt} = /_n K_s C_s - \frac{Tm_a Y}{Y_a}\left(\frac{C_{aq}}{K_s + C_{aq}}\right) C_{xa} \qquad \text{[E 5]}$$

Dynamic biogas model:
Acidogenic Biomass:

$$\frac{dC_{xa}}{dt} = (\frac{m_a}{Y_a})(\frac{C_{aq}}{K_s + C_{aq}} - K_{da}) C_{xa} \qquad [E\ 6]$$

Methanogenic Biomass:

$$\frac{dC_m}{dt} = \frac{T m_m Y}{Y_m}(\frac{C_{ac}}{K_s + C_{ac}} - K_{dm}) C_m \qquad [E\ 7]$$

Acetate:

$$\frac{dC_{ac}}{dt} = Y_{ac}((1 - Y_a)(\frac{m_a}{Y_a})(\frac{C_{aq}}{K_s + C_{aq}} + K_{ds}) C_{xa} - (\frac{m_m}{Y_m})(\frac{C_{ac}}{K_s + C_{ac}}) C_{xm} \qquad [E\ 8]$$

Methane:

$$\frac{dC_{CH4}}{dt} = Y_{CH4}((1 - Y_m)(\frac{m_m}{Y_a})(\frac{C_{aq}}{K_s + C_{aq}} + K_{da}) C_{xa} \qquad [E\ 9]$$

Carbon Dioxide:

$$\frac{dC_{CO2}}{dt} = (1 - Y_{CO2})((1 - Y_a)(\frac{m_a}{Y_a})(\frac{C_{aq}}{K_s + C_{aq}} + K_{da}) C_{xa} + \qquad [E\ 10]$$
$$(1 - Y_{CH4})((1 - Y_m)(\frac{m_m}{Y_m})(\frac{C_{ac}}{K_s + C_{ac}} + K_{dm}) C_{xm}$$

This simple dynamic model gives a first inside in the generation of methane and carbon dioxide.

The biogenic part describing the microbial and the growing equation are given by the following parameters: K_s, K_d, μ und Y, describing the microbial fermentation processes.

The other part of the model describes the conversion of the solid carbon Cs in solved aqueous organic carbon C_{aq}. The aqueous carbon C_{aq} is converted into acidogenic carbon C_{xa} and in acetate C_{ac}. The carbon C_{ac} is converted into C_{xm} methanogenic carbon. The generation of methane C_{CH4} is done from C_{xm}. The generation of C_{co2} is based on C_{xm} and C_{xa}. This is the simple flow diagram described by the set of differential equations. [24]

In the following part some basic simple chemical reaction is listed up which can occur during the anaerobe fermentation. The process starts with cellulose. The cellulose is converted into glucose. Glucose is converted into carboxylic acids (acetic, propionic, butyric). The acetic acids are the basis for methane and carbon dioxide generation, additional from hydrogen and carbon dioxide methane and water are generated. These simple reactions show a simple system of chemical reactions inside the biogas plant.

Cellulose: $C_6H_{10}O_5 + H_2O \longrightarrow C_6H_{12}O_6$
Glu cos e: $C_6H_{10}O_6 \longrightarrow CH_3(CH_2)_2COOH + 2H_2 + 2CO_2 + Q(+\ 197kJ/mol)$
Glu cos e: $C_6H_{10}O_6 + 2H_2 \longrightarrow 2CH_3CH_2COOH + H_2O + Q(+\ 216kJ/mol)$ [E 11]
Glu cos e: $C_6H_{10}O_6 + 2H_2O \longrightarrow 2CH_3COOH + 4H_2 + 2CO_2 + Q(+\ 164kJ/mol)$
Butyric acid: $CH_3(CH_2)_2COOH + 2H_2O \longrightarrow 2CH_3COOH + 2H_2 + Q(-\ 34kJ/mol)$
Propionic acid: $CH_3CH_2COOH + 2H_2O \longrightarrow 2CH_3COOH + 3H_2 + CO_2 + Q(-\ 25kJ/mol)$
Acetic acid: $CH_3COOH \longrightarrow CH_4 + CO_2 + Q(+\ 16kJ/mol)$
Hydrogen: $4H_2 + CO_2 \longrightarrow CH_4 + 2H_2O + Q(+\ 31kJ/mol)$

The physical Process describing the diffusion process inside the biogas plant from the solid parts to the gaseous and liquid parts, and inside the gaseous and liquid parts are in reality not known. For every simulation we need the distribution of the solid parts and the sizes and the material types. In normal we know nothing about these parameters. Because of the lack of information on the distribution of the solid parts, also the heat conduction inside the biogas plant is not known. Only measurements of the temperature give an average temperature inside the local biogas plant structure. Diffusion and heat conduction cannot be simulated in detail because of the lack of data and information. [24[25]

Based on the equation of the simple dynamical model measurement in the temperature distribution, measurements in the gas volume flow and gas components mixture, enables to use the evaluation model according to the VDI 2180: 2015-1. [22]. With the model equation the measured parameters can be evaluated, whether they are confident or no, whether the model predicts in the right way. To start the evaluation the following empirical parameter can be used to start the simulation:

	μ_m	Y_m	K_{sm}	K_{dm}		μ_a	Y_a	K_{sa}	K_{da}
Min	0,1	0,05	10	0,01		2	01,	10	0,01
Aver	0,25	0,06	500	0,03		10	0,15	50	0,05
max	0,5	0,07	2500	0,04		30	0,5	150	0,4

Table 1: Table of measured and empirical values of the bio kinetic parameters for the generation of methane and acetogens. (Source: [24])

The advantage in using a set of nonlinear time dependent differential equation is the set of parameters and variables, which can be compared with the measurements data done on plant. Now we see the advantage of a simple set of equations compared with an empirical assumption and measurement of the biogas plant. The general form of the above equations is given by

$$\frac{dx}{dt} = f(x,t,a) \qquad \text{[E 12]}$$

If the set of equations possesses a solution we can write

$$x = x_0 + v \ \& \ \frac{dx_0}{dt} + \frac{dv}{dt} = f(x_0 + v, t, a) = (f(x_0, t, a) + \frac{2f}{2x}v + O(v^2)) \qquad \text{[E 13]}$$

Linear System by evaluating the Jacobian Matrix **F**:

$$\frac{dv}{dt} = \frac{2f}{2x} v = f(x) + Fv = 0 \quad \& \quad v = F^{-1} f(x) \qquad \text{[E 14]}$$

Now we can apply the idea of solving a nonlinear set of equations with measurement values and probabilistic deviation values. We combine the stationary nonlinear set of equations and the boundary conditions (BC):

$$\text{BC:} \quad f(x, a) = 0 \qquad h(x) = 0 \qquad \text{[E 15]}$$

The measurement leads to the deviation of the exact solution x and the measurement μ:

$$x = m - dx \qquad \text{[E 16]}$$

In the probabilistic theory we have the expected value E(x) and the variance Var(x) and the covariance of the values x: Cov(x):

$$m = E(x) \qquad s_x^2 = Var(x) = E(x - E(x))^2$$
$$r(x_i, x_k) = \frac{Cov(x_i, x_k)}{\sqrt{Var(x_i) Var(x_k)}} \qquad Cov(x_i, x_k) = E(x - m_i)(x_k - m_k) \qquad \text{[E 17]}$$

With this basis we can now calculate the confidential values and detect the failures and fault in the measurements by solving a minimizing problem with Lagrange factor: [22]

$$h(x) = h(m - a) = 0$$
$$z = a S_x^{-1} a - 2(h(x) + \frac{2h}{2x} a) l \qquad \frac{2z}{2a} = 0 \qquad \text{[E 18]}$$

The solution of the minimizing problem is given by the Lagrange factor λ, deviation vector a, the Covariance matrix $\mathbf{S}_v$

$$l = -((\frac{2h}{2x}) S_x (\frac{2h}{2x})^T)^{-1} h(x)$$
$$a = -S_x (\frac{2h}{2x})^T ((\frac{2h}{2x}) S_x (\frac{2h}{2x})^T)^{-1} h(x) \qquad \text{[E 19]}$$
$$S_v = (\frac{2a}{2x}) S_x (\frac{2a}{2x})^T$$

The empirical Covariance matrix $\mathbf{S}_X$ is calculated from the measurements and the boundary conditions. Applying the described method from above, we can calculate the deviation vector a, which shows the new variable values and parameter values and the changing of the values during the operation. [22]

A detailed description is given in the VDI 2048-1:2015. It looks very complicated, but in practical it is very east to be applied at every dynamical system, and it enables to verify the reliability of the measurements and the assumed values in the dynamical system and process. [22]

If we compare the empirical model Biogas Plant with the simple set of differential equation shown above has the advantage to combine measurements with physical and chemical models, we additional observe that the empirical model can be applied to nearly all biogas plant. For biogas plants modelled by a bioreactor, the coupled set of dynamic equations can be applied.

The result of the coupled dynamic equations enables to simulate the methane generation, the carbon dioxide generation, as a function of solid carbon inside the biogas plant. With water the solid carbon will be solved as aqueous solved carbon. The water content and the distribution inside the biogas plant is important. Additional we have to recognize that biogenic biomass type and the distribution of the biogenic biomass is needed.

This simple model can be applied to the VDI 2048-1:2015 a standard for validation of the process data in combination with the measurement of the process data. The measurement and the calculation show the confidentiality of the assumed parameters and constants. The procedure of the VDI 2048-1:2015 can also be applied to the biogas plant model. But the information gained is small, because the generation of biogas is the only data to be checked. [22]

With the calculated and measured biogas production the next steps in the conversion of biogas can be analyzed: gas cleaning as a function of measurement of pollution and additives. After gas cleaning the step gas preparation and biogas conversion follow. In the next chapter gas cleaning and gas preparation are described and compared.

1.7 Combustion biogas in gas engine

The combustion of biogas after the cleaning process is the classical application producing electric energy and thermal heat, the well-known combined heat process (CHP).

1.7.1 Gas Engine

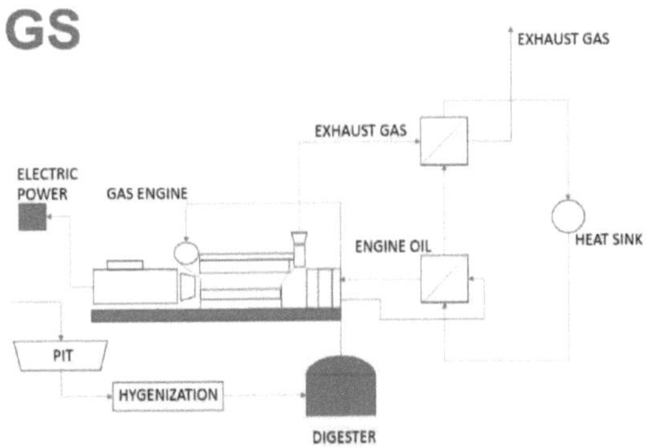

Figure 2: Combustion of biogas in well-known gas engines electric power range Pele = 500 kW_{ele} up to 2000 kW_{ele}, electric efficiency η_{ele} = 32% up 40%, biogas with methane concentration x CH4 = 40% up to 65%. (Source: [2])

The combustion of biogas in a gas engine is another conversion to heat and electricity. The gas engine is supporting a generator. The efficiency of the gas engine is given for biogas (lean gas) η_{ele} ~ 32% up to 40% and thermal efficiency η_{th}~40% up to 45%. The CHP solution is well known and applied in many projects. The main part is the combustion of cleaned biogas gas. The electric efficiency the gas engine η_{ele} = 32% up to 40%. The combustion of biogas in the CHP application for generation of electricity and heat is well-known. Electricity defines the economic part. The economic efficiency is lower than gas upgrading and much lower than liquefying of biogas to DME.

Some useful relations are mentioned: [24]

Substrate (FM)	P_{ele}	Biogas(m^3/h)	Biogas (CH_4/CO_2)
— 17 500 t/a	625kWh	~350m^3/h	Biogas (50%/50%)
— 35 000 t/a	1300kWh	~700m^3/h	Biogas (50%/50%)

Later on, we will come back to the gas engine generating electric power and thermal power needed in the biogas conversion process.

1.8 Take Over Biogas

Till this point we have described and modelled and analyzed the classical biogas plant. The model shows the dynamical behavior of the biogas plant as a simplified reactor. The biogas the product of the fermentation process.

In most application we take over the biogas. Taking over means: measurement of the volume flow, the measurement of pressure and temperature, the measurement of the methane concentration, carbon dioxide concentration, measurement of the water, hydrogen sulfur concentration and ammonia concentration.

Taking over biogas is the same as taking over a raw gas, which has to be cleaned, prepared to be used for the next process steps.

1.8.1 Low pressure compressor

The function of the low-pressure compressor is to suck the biogas from the digester

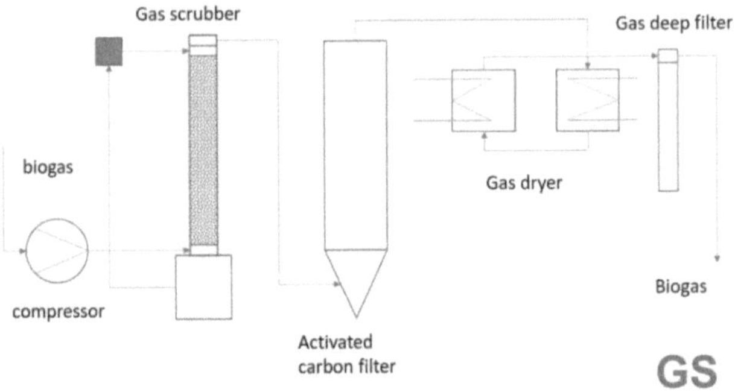

and to press the biogas over the biogas cleaning route:

Figure 3: The drawing shows the biogas cleaning route. (Source: [2])

1.8.2 Biogas Preparation

Biogas preparation includes the gas cleaning and in the second step gas separation, in special gas upgrading concentrated on methane. Today used processes are chemical scrubber with e.g. amine, water, potassium, glycol, pressurized water systems. The simplest cleaning of the biogas can be done by using activated carbon with pellets. Besides filtering gas dryers are used. If filtering is used a fine filter filtering particle dp ~ 0,001µm.

1.8.3 Biogas Scrubber

As wash fluid are used: water, biodiesel. The scrubber consists of a cylindrical structure filled with washing parts, on the surface the gas is contacting the wash fluid and the absorption process takes place. To support the absorption process packing materials, mainly rings and structures, are used inside the cylindrical column. The wash fluid is running over the surface of the packing structures and the gas flowing over the wet surface. The packing structures enable to enlarge the contact surface. The saturated wash fluid is regenerated and used again. The wash process is a closed cycle absorption and regeneration. In the most cases the regeneration is done in vacuum, with a stripping fluid and heat.

1.8.4 Activated Carbon Filter

Activated carbon is pressed to pellets, which are used in gas cleaning to reduce pollutions like H2S, VOC in biogas gas. The principle is simple: because of the large surface of carbon and the porous structure activated carbon is used for absorption and adsorption of molecules. The pellets are located inside a cylindrical tank, and the gas is streaming through the fixed bed of pellets. Measurement of the gas concentration inlet and outlet, shows the filtering capacity and the time to change the pellets for new ones.

1.8.5 Gas Drying

Under Gas drying we mean the reduction of water in the biogas gas, described by the dew point. By compression of the biogas the pressure is increased and the gas will be cooled down. Using the cold water in the temperature range 1°C up to 4°C, the water is removed as condensate. Using a refrigeration fluid (combination of dimethyl ether and carbon dioxide = blend) we can cool down the gas under 0°C but this is difficult ice can be built. Cooling down the biogas is described by the dew point temperature (T_d) as a function of temperature in combination of the relative humidity (f (T=°C)):

$$Td = \left(\frac{f}{100}\right)^{\frac{1}{8}}(112 + 0.9T) + 0.1T - 112 \qquad \text{[E 20]}$$

1.9 Deep Gas Filter

To reduce the dust and to clean up the biogas a deep multistage filtering is used. The filter membrane is made of synthetic cloth stabilized with a stainless rod grid, so that the membrane cannot collapse. To clean up a dust with the diameter d~0.001µm we use a two-stage membrane filtering system.

At this point the biogas is cleaned up to be converted into a synthetic gas, gas mixture of carbon monoxide (CO), hydrogen (H_2), carbon dioxide (CO_2).

1.10 Biogas Conversion to Dimethyl ether (DME)

After biogas cleaning we focus on the next step, the conversion of biogas to dimethyl ether. In the next passages we describe several possibilities of liquefaction of biogas to dimethyl ether (DME). [1, [2].

Dimethyl ether is well-known. It is the simplest ether generated from methanol by dehydration. To realize the methanol-synthesis we need synthetic gas.

The first step of liquefaction is to generate a gas mixture called synthetic gas consisting of carbon monoxide (CO), carbon dioxide (CO_2) and hydrogen (H_2). In the second step a conversion of the synthetic gas mixture to dimethyl ether is done.

Because we have cleaned biogas there are two pathways to generate synthetic gas, the first way using dry reforming, the second way using water steam reforming.

1.10.1 Dry Reforming

Dry reforming is the conversion of methane in combination with carbon dioxide to carbon monoxide, hydrogen and carbon dioxide. Besides dry reforming steam reforming of carbon monoxide (Water gas shift reaction = WGS) is applied. This reduces the production of solid carbon: [6],[11],[12]

$$CH_4 + CO_2 \longrightarrow 2CO + 2H_2 + Q(+\ 247kJ/mol) \qquad [E\ 21]$$
$$CO + H_2O \longrightarrow CO_2 + H_2 + Q(+\ 41.2kJ/mol)$$

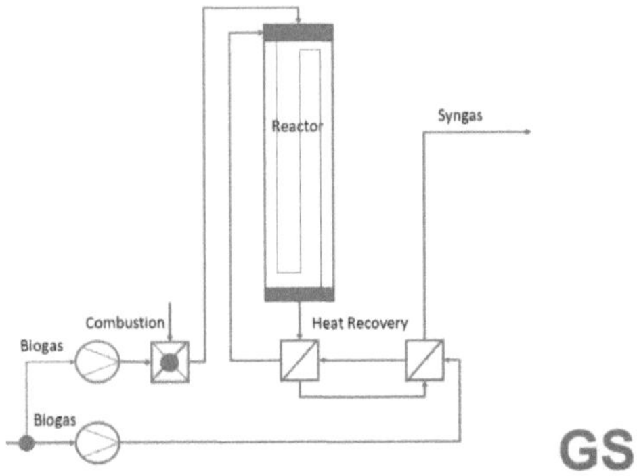

Figure 4: Dry Reforming Reactor. In most cases dry reforming is realized in combination of the water gas shift reaction and several catalysts like Nickel, Calcium-oxide, Iron-oxide. (Source: [20])

The dry reforming process operates at temperature T~800°C up to 1000°C and has a conversion rate supported by catalyst up to 98%. Dry Reforming process is strong endothermic under application of catalysts the operation temperature in the reactor is given T ~ 700°C.

The catalysts are based on Ni basis in combination with Calcium oxide and Iron oxides. Lowering the operating temperature T ~ 600°C is one of the aims to be reached by using catalysts. The lower operation temperature enables to use standards materials to lower the costs of material, the increase the stability and efficiency.

One problem of dry reforming is the production of solid carbon. This carbon is not desired. Therefor dry reforming is combined with additional water steam. The chemical reactions are:

$$CO + H2 \longrightarrow C + H2O + Q(-\ 135kJ/mol) \quad (K1)$$
$$C + CO2 \longrightarrow 2CO + Q(+\ 180kJ/mol) \quad (K2)$$
$$CO2 + CH4 \longrightarrow 2CO + 2H2 + Q(+\ 251kJ/mol) \quad (K3) \qquad [E\ 22]$$
$$C + 2H2 \longrightarrow CH4 + Q(-\ 71kJ/mol) \quad (K4)$$
$$CO + H2O \longrightarrow CO2 + H2 + Q(-\ 45kJ/mol) \quad (K5)$$

The operating pressure is in most cases near the environment pressure p~ 1,5 bar, so the equilibrium constants depend strong on the reactor temperature. The equilibrium constants for the chemical reaction (K1...K5) mentioned above are shown as a function of temperature:

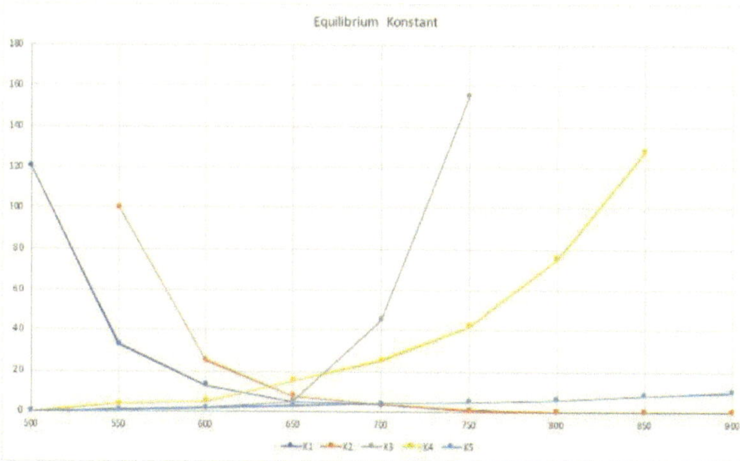

Figure 5: Dry Reforming process reactions Equilibrium coefficient of the different reactions as a function of temperature. The temperature range is given between 500°C up to 850°C. (Source: [23])

One problem of dry reforming is the production of solid carbon. This carbon is not desired. Therefor dry reforming is combined with additional water steam. The combination of steam reforming and dry reforming is stable and reduces solid carbon to carbon monoxide and hydrogen by the additional water. Operation Pressure p ~1.5 bara and the operating temperature in the reactor T~ 850°C up to 900°C, CO2: H2O = 1,0:0,4 up to 0,6:1,5. [30]

$$CH_4 + CO_2 \longrightarrow 2CO + H_2 + Q(+\ 247kJ/mol) \qquad [E\ 23]$$
$$CH_4 + H_2O \longrightarrow CO + 3H_2 + Q(+\ 206kJ/mol)$$

The chemical reaction system is strong endothermic.

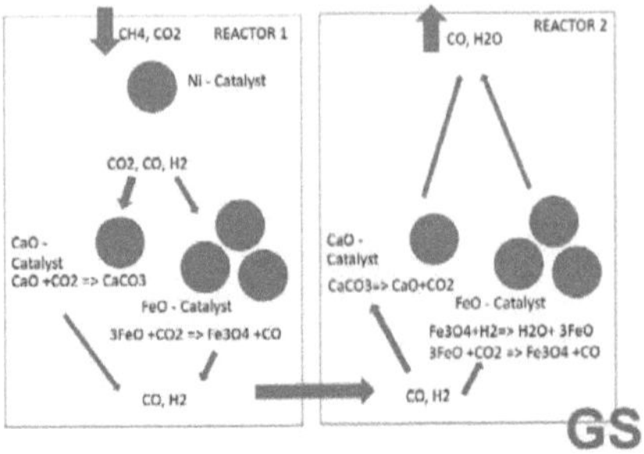

Figure 6: Super Dry reforming with a combination of three catalysts, Nickel catalyst, Calcium oxide and Iron oxide enables to convert methane (CH4) and carbon dioxide (CO2) to carbon monoxide (CO) and steam (H2O). This concept can be enlarged and is strong controllable. (Source: [30])

Dry reforming is a new field in chemical engineering searching for possibilities of reduction of carbon dioxide (CO_2). The interest in dry reforming is the conversion of methane and carbon dioxide to carbon monoxide and hydrogen. Dry reforming supported by more than here catalysts is understood by the following sum reaction:

$$CH_4 + 3CO_2 \longrightarrow 4CO + 2H_2O + Q(+\ 341 kJ/mol) \qquad [E\ 24]$$

To reduce the heat added to the reactor for the dry reforming, heat recovery at dry reforming is used. The inlet gas (CO_2+CH_4) or (CH_4+3CO_2) is heated up to T~ 600°C, so that the temperature difference in the reactor is given in the range of dT ~ 200°C. The heat for the highly endothermic reaction can be generated by methane (auto thermal) or by external generation with ceramic heat tubes converting electric energy to heat (small units).
Heat recovery at dry reforming. Under heat recovery we understand using the latent heat of the exhaust gas of the reactor in heating up the inlet gas. This reduces the needed energy in the dry reforming reactor.

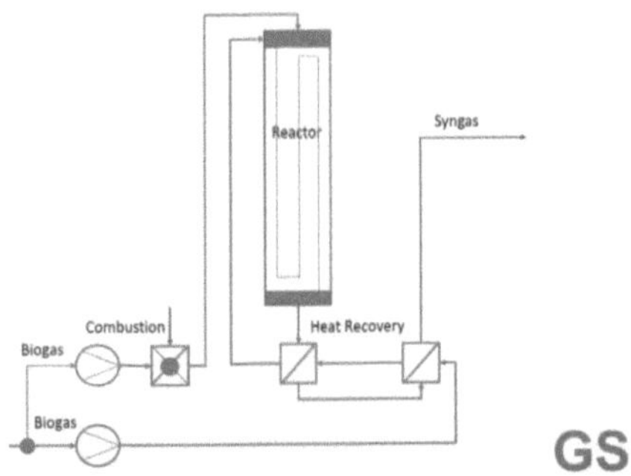

Figure 7: Dry reforming in combination with chemical looping. Under chemical looping we understand oxidizing of a metal in the dry reforming reactor, deoxidizing in a separate reactor under heat supply. If the heat is generated by electricity we get a clean oxygen gas. If we generate heat from oxidizing methane we get carbon dioxide and steam. (Source: [30])

The idea of using more than one catalyst is a modified application of chemical looping is using the property of oxidation of metals at high temperatures and the regeneration of metal oxides at high temperatures.

$$Me + O \longrightarrow MeO \qquad [E\ 25]$$
$$MeO \longrightarrow Me + O$$

For calcium oxide the chemical reaction is given:

$$CaO + CO2 \longrightarrow CaCO3 + Q(-\ 172kJ/mol) \qquad [E\ 26]$$
$$CaCO3 \longrightarrow CaO + CO2 + Q(+\ 172kJ/mol)$$

For iron oxide the chemical reaction is given:

$$Fe_3O_4 + H2 \longrightarrow 3FeO + H_2O + Q(+\ 79kJ/mol) \qquad [E\ 27]$$
$$3FeO + CO2 \longrightarrow Fe_3O_4 + CO + Q(-\ 34kJ/mol)$$

The combination of three catalysts enable to convert more than 3 CO_2 with methane (CH_4):

$$CH_4 + xCO_2 \longrightarrow (\frac{x}{2} + 1)CO + 2H_2O + (\frac{x}{2} - 1)\frac{1}{2}O_2 \qquad [E\ 28]$$

This shows that the additional oxygen can be fixed with the iron oxide. The regeneration can be done by using heat and hydrogen converted to steam. In the practical realization of dry reforming we have to add steam (H_2O) and we will observe additional reactions:

$$CH_4 + CO_2 \longrightarrow 2C + H_2O + Q(-\ 17kJ/mol)$$
$$C + H_2O \longrightarrow CO + H_2 + Q(-\ 290kJ/mol) \qquad [E\ 29]$$
$$CO + H_2O \longrightarrow CO_2 + H_2 + Q(-\ 45kJ/mol)$$

Dry reforming leads to a synthetic gas mainly consisting of carbon monoxide, steam and small amount of methane and carbon dioxide and hydrogen.

Beside the pathway of dry reforming steam reforming can be applied to biogas to generate synthetic gas.

1.10.2 Steam Reforming

Steam Reforming is wide spread over, used in generation of hydrogen and carbon dioxide in combination with the water gas shift reaction (WGS). It is very simple process and stable. [3]

$$CH_4 + H_2O \longrightarrow CO + 3H_2 + Q(+\ 206kJ/mol) \qquad [E\ 30]$$
$$CO + H_2O \longrightarrow CO_2 + H_2 + Q(-\ 41.2kJ/mol)$$

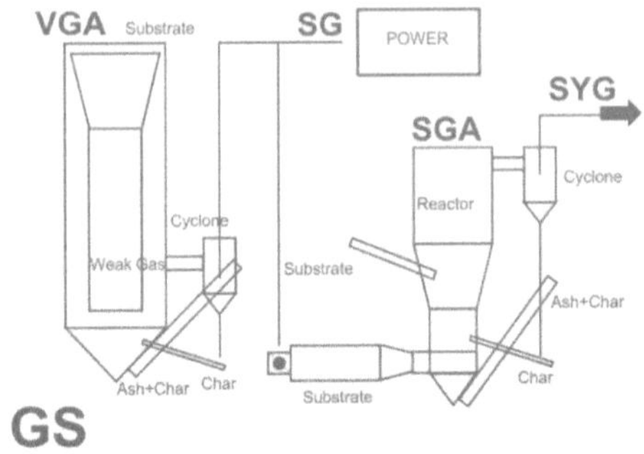

Figure 8: Drawing shows the steam gasification as realized in the project Traismauer [1] and Hagenbrunn [2], based on a top cycle (steam gasification) and a bottom cycle (gasification). (Source: [1])

Steam reforming generates a gas mixture on carbon monoxide, carbon dioxide and hydrogen, depending on steam the amount of carbon monoxide can be reduced to several hundred ppm ~ 500 ppm.

Primary reformer:

- Overall heat of reaction is endothermic
- multi-tube reformer
- Reactions are supported over Ni-catalyst
- Reactions at the pressure of 15 up to 30 bara and in combination at a temperature T~850°C up to 950°C range,
- $H_2/CO > 3$

Carbon formation:

$$2CO \longrightarrow C + CO_2 + Q(-\ 173kJ/mol)$$
$$CH_4 \longrightarrow C + 2H_2 + Q(+\ 75kJ/mol) \qquad [E\ 31]$$
$$CO + H_2 \longrightarrow C + H_2O + Q(-\ 132kJ/mol)$$

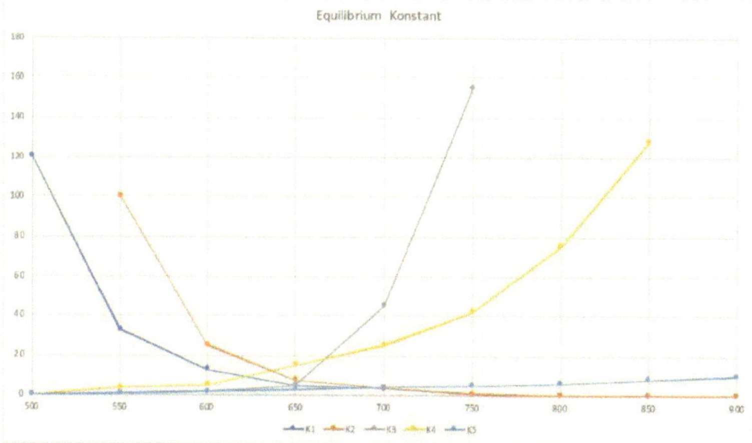

Figure 9: equilibrium constants for the chemical reactions of steam reforming. The equations of [30] are calculated and plotted in a temperature range of T=500°C up to 900°C, and a pressure range p ~ 1.5 bar (nearly environment pressure). (Source: [35])

Carbon deposits deactivates the catalyst. Actions to reduce carbon formation:

- High steam/carbon (S/C) ratio reduces carbon formation.
- Expensive to produce steam.
- Addition of CO_2 reduces carbon formation
- Pre-reformer if higher hydrocarbons are present
- Common S/C-ratio is 2.5–4.5

1.10.3 Methanol Synthesis und Dehydration

There are two methods in liquefaction the syngas mixture of carbon dioxide, hydrogen and carbon monoxide. One method is applying in the first step the methanol synthesis and in the second step the dehydration to dimethyl ether. The other method is the direct synthesis of syngas mixture to dimethyl ether in the reactor under the combination of two catalysts. [3],[13],[14],[15],[30].

Looking at the method of methanol synthesis we have two possibilities: conversion of carbon monoxide and hydrogen to methanol, conversion of carbon dioxide and hydrogen to methanol. Both reactions are possible. Analyzing the chemically reactions the conversion of carbon dioxide and hydrogen to methanol is very fast and stable. During the generation of methanol water (steam) is produced. Steam has to be reduced as much as possible because steam is prohibiting the conversion process on the catalyst surface.

$$CO + 2H_2 \longrightarrow CH_3OH + Q(-\ 171kJ/mol) \quad [E\ 32]$$

$$CO_2 + 3H_2 \longrightarrow CH_3OH + H_2O + Q(-\ 118kJ/mol)$$

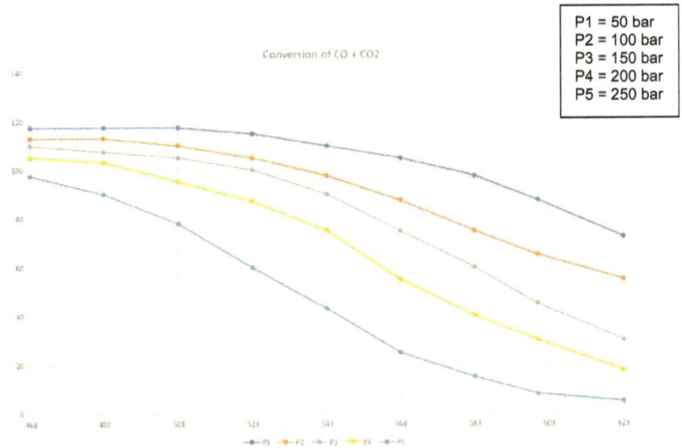

Figure 10: Low Pressure Methanol Synthesis as a function of pressure and temperature and the conversion of carbon monoxide (CO) and carbon dioxide (CO2) to methanol. The temperature range is given from T=190°C up to 350°C, pressure range from 50 bar up to 250 bar. In the biogas and biomass application we operate in the low-pressure range p=50 bar up to 100 bar to reduce the plant costs. (Source: [35])

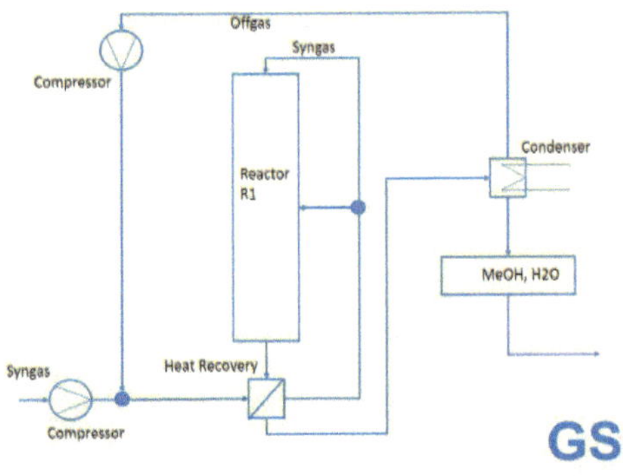

Figure 11: The conversion of syngas to methanol in a reactor depends on the conversion rate (methanol yield). The heat of the product steam is recovered by the fresh input syngas. The offgas is gained after the condensation of the product steam mainly consisting of methanol and water. The offgas consist of hydrogen, carbon monoxide, carbon dioxide and light hydrocarbons. The offgas is recycled in the reactor again. (Source: [2])

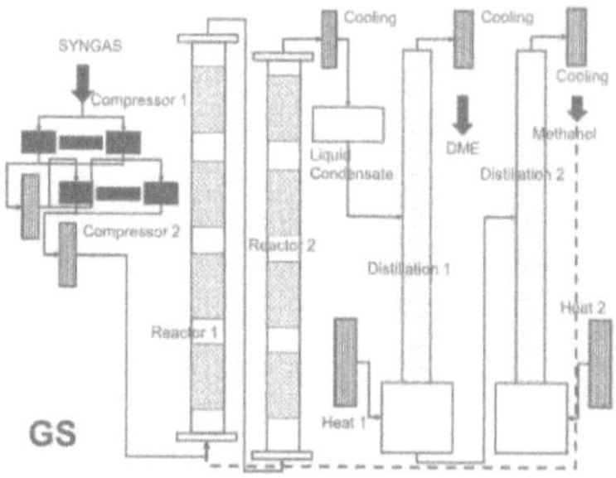

Figure 12: Process of methanol synthesis – George Olah [30] applied to Traismauer [1] and Hagenbrunn [2]- this is a simple process flow, for the methanol reactor we have to apply the heat recovery and recycling of the offgas. (Source: [2])

The generation of methanol on the basis of CO_2 and H_2 is done in one reactor. The generated steam mixture of water steam and methanol is cooled down and the pressure reduced. The condensate is a mixture on water and methanol and dimethyl ether. The separation is done in two steps. First the reduction of dimethyl ether and second the reduction of water.

$$\begin{aligned} &CO + H_2O \longrightarrow CO_2 + H_2 + Q(-\ 41kJ/mol) \\ &CO_2 + 3H_2 \longrightarrow CH_3OH + H_2O + Q(-\ 118kJ/mol) \\ &2CH_3OH \longrightarrow CH_3OCH_3 + H_2O + Q(-\ 23.5kJ/mol) \end{aligned} \qquad [E\ 33]$$

The efficiency of the process based on CO_2 is given by 87% at a pressure of p ~ 50 bara and T ~ 300°C catalysts used: CuO/ ZnO /γ-Al_2O_3.

An alternate way of generation of methanol is from carbon monoxide and steam using the WGS:

$$\begin{aligned} &CO + H_2O \longrightarrow CO_2 + H_2 + Q(-\ 41kJ/mol) \\ &CO + 2H_2 \longrightarrow CH_3OH + Q(-\ 171kJ/mol) \\ &2CH_3OH \longrightarrow CH_3OCH_3 + H_2O + Q(-\ 23.5kJ/mol) \end{aligned} \qquad [E\ 34]$$

The efficiency of the process based on CO_2 is given by 92% at a pressure of p ~ 50 bar and T ~ 300°C catalysts used: CuO/ ZnO / γ-Al_2O_3.

The methanol synthesis is an exothermic reaction depending on temperature and pressure. We are only focusing on pressure in the range p~ 30 bara up to 50 bara.

We learned from the measurement and analysis that the reactions based on carbon dioxide is faster than carbon monoxide. In practice we always have a gas mixture on CO_2 and CO and H_2. Therefor we always have a combination in methanol synthesis of WGS by carbon monoxide and Carbon dioxide in combination with hydrogen. This deviation from the ideal shows the combination of carbon dioxide and carbon monoxide.

$$CO_2 + 3H_2 \longrightarrow CH_3OH + H_2O + Q(-\ 49.8 kJ/mol)$$
$$CO + 2H_2 \longrightarrow CH_3OH + Q(-\ 91.0 kJ/mol) \qquad [E\ 35]$$
$$CO_2 + H_2 \longrightarrow CO + H_2O + Q(+\ 41.2 kJ/mol)$$

The methanol syntheses depend on temperature and pressure as shown in the following drawings:

Figure 13: Methanol synthesis as a function on pressure and temperature, pressure range p=10 bar up to 100 bar, temperature from T=130°C up to 330°C. (Source: [20])

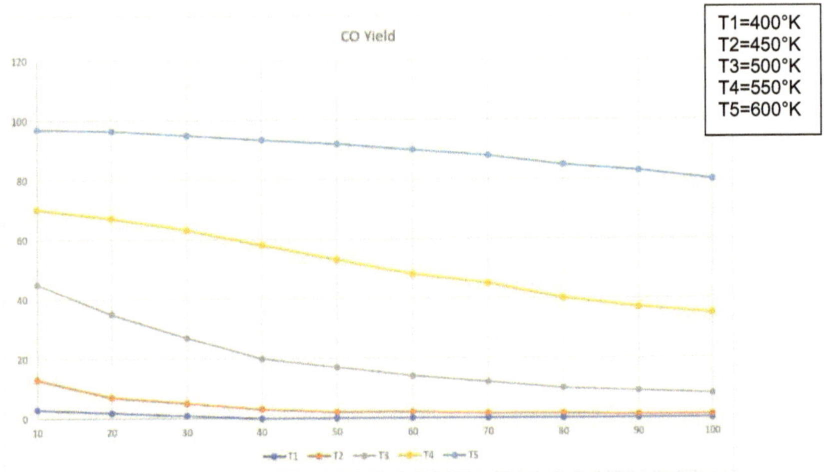

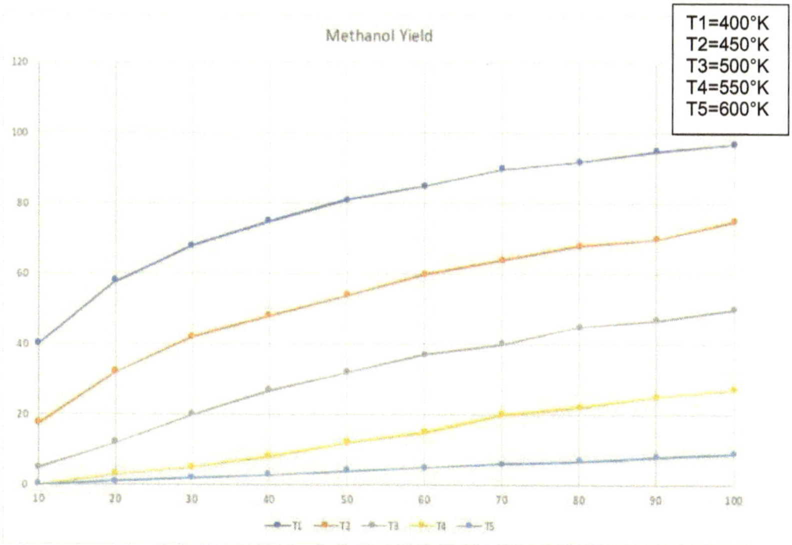

Figure 14: Methanol synthesis as a function on pressure and temperature, pressure range p=10 bar up to 100 bar, temperature from T=130°C up to 330°C. (Source: [20])

1.10.4 Summary pathways to dimethyl ether

1.10.5 Two-step process toward DME

The two-step process is based on the generation of methanol stored in a tank. From Methanol we can generate by dehydration dimethyl ether. The chemical reaction is given by:

$$2CH_3OH \longrightarrow CH_3OCH_3 + H_2O + Q(+\ 47.4kJ/mol) \qquad [E\ 36]$$

We use two different catalysts:

- Reactor 1: Methanol syntheses: CuO-ZnO-γ-Al_2O_3 (CO preferred, H_2)
- Reactor 2: Dehydration: HSZM-5

or

- Reactor 1: Methanol syntheses: CuO-ZnO-γ-Al_2O_3 (CO_2 preferred, H_2)
- Reactor 2: Dehydration: HSZM-5

The result is liquefaction biogas in form of DME stored in tanks at pressure of p~15 bara in liquid phase.

1.10.6 One-step process toward DME

The one-step method is called the direct synthesis and done in one reactor. The chemical reactions are given:

$$2CO + 4H_2 \longrightarrow 2CH_3OH + Q(-\ 256kJ/mol)$$
$$CO + H_2O \longrightarrow CO_2 + H_2 + Q(-\ 45kJ/mol) \qquad [E\ 37]$$
$$2CH_3OH \longrightarrow CH_3OCH_3 + H_2O + Q(47kJ/mol)$$

- -

$$3CO + 3H_2 \longrightarrow CH_3OCH_3 + CO_2 + Q(-\ 254kJ/mol)$$

The reaction velocity in the reactors are very low, the resting time inside the reactor has to be t ~ 170 sec up top 250 sec. The reaction is exothermal, so that heat is generated, which can be used for the distillation of the condensate of methanol, water and dimethyl ether. Looking at the reaction above, we can see that carbon dioxide is generated. Therefor it is not a real carbon dioxide sink:

We use two different catalysts:

- Reactor 1: Methanol syntheses: CuO-ZnO-γ-Al_2O_3 (CO preferred, H_2)
- Reactor 1: Dehydration: HSZM-5

or

- Reactor 1: Methanol syntheses: CuO-ZnO-γ-Al_2O_3 (CO_2 preferred, H_2)
- Reactor 1: Dehydration: HSZM-5

$$2CO_2 + 6H_2 \longrightarrow 2CH_3OH + 2H_2O + Q(-\ 166kJ/mol)$$
$$2CH_3OH \longrightarrow CH_3OCH_3 + H_2O + Q(-\ 47kJ/mol) \qquad [E\ 38]$$

- -

$$2CO_2 + 6H_2 \longrightarrow CH_3OCH_3 + 3H_2O + Q(-\ 119kJ/mol)$$

The direct synthesis based on CO_2 has a higher conversion velocity. The reaction velocity in the reactors are very low, the residence time inside the reactor has to be t ~ 170 sec up top 250 sec. The reaction is exothermal, so that heat is generated, which can be used for the distillation of the condensate of methanol, water and dimethyl ether. Looking at the reaction above we can see that steam (water) is generated. During the process steam has to be removed from the surface of the catalyst. Steam is acting prohibiting. One very interesting property is that no carbon dioxide will be generated, therefore we have a real CO_2 sink. [30]

1.10.7 Biogas conversion to DME

The conversion of biogas to dimethyl ether leads to some practical relations based on the measurements in the project Hagenbrunn [2]:

BIOGAS							
		m³/h	t/h				
P(kW ele)	Qzu(kWh)	Biogas(5.0kWh/m³)	Substrat	t/Jahr	Peigen(kWh)	DME (l/H)	Pele(kWh)DME
50.00	166.67	33.33	0.22	1777.78	10.00	16.67	8.33
100.00	333.33	66.67	0.44	3555.56	20.00	33.33	16.67
150.00	500.00	100.00	0.67	5333.33	30.00	50.00	25.00
200.00	666.67	133.33	0.89	7111.11	40.00	66.67	33.33
250.00	833.33	166.67	1.11	8888.89	50.00	83.33	41.67
500.00	1666.67	333.33	2.22	17777.78	100.00	166.67	83.33
1000.00	3333.33	666.67	4.44	35555.56	200.00	333.33	166.67

1l Diesel = 3.09 kg CO2 1m³ Methan = 2.99 kg CO2

Table 2: Conversion of biogas to dimethyl ether. This table enables to get a better feeling on the relations of biogas, generated electric power, needed own electric power and produced DME (l/h). (Source: Johann Gruber-Schmidt, 2016)

1.11 Dimethyl ether (DME)

The conversion of biogas to dimethyl ether (DME) can be interpreted as liquefaction of biogas gas. DME is the simplest ether, then chemical formula is CH3OCH3, two methyl groups are connected over an oxygen atom. The LHV = 8,3 kJ/L, DME is liquid at a pressure p =6 bara and T=25°C and can be stored in a tank. Thermodynamic data and properties can be taken from the database NIST. There are several pathways to synthesize DME from syngas. Syngas is a gas mixture of carbon dioxide (CO_2), carbon monoxide (CO) and hydrogen (H_2). The combustion of DME in a standard diesel engine has been studied 2014 in the USA on the road by heavy trucks. 2013 the IDA (USA) has DME certified as international fuel by the ISO 16 618: 2013. Additional the ASTM 7905: 14A and ASTM 7905:14B can be used. DME can be used as a substitute for fossil diesel. Beside the application as a fuel, DME can be used as a substitute for LPG. In the chemical industry DME is often used as a propellant to produce expanded hard polystyrene, an insulation and construction material. Dimethyl ether is also a good hydrogen storage fluid, which can be used in combination with fuel cells, realization of zero emission and high electric efficiency.

The advantages are listed up:

- Liquid fuel simple handling and storing
- substitute of LPG (liquid propane gas)
- very good combustion properties (very small amount of CO, NO_x, no soot and no fine dust)
- higher economic success
- high amount of hydrogen
- generation from CO_2 and H_2
- substitute of fossil Diesel
- simple storage
- supporting zero emission
- simple transport

Companies like VOLVO, Heavy Trucks, FORD or MACK TRUCK sell commercial trucks running on DME and have several years proofed successful DME on the road. [32],[33].

1.12 Reduction of Greenhouse gas (GHG)

Using dimethyl ether (DME) instead of fossil Diesel is a first step in reduction of greenhouse gas (GHG) in the exhaust gas, and the reduction of soot, nitrogen oxide and dust. A further development is the property of zero emission. Zero Emission DME is generated from CO_2 and H_2O. Converting dimethyl ether (DME) to electricity and heat, the resulting exhaust gas is steam mixture on carbon dioxide and steam, which can be separated in CO_2 and H_2O. The generation of DME from CO_2 and H_2O is called regenerative DME ((Reg)DME). The generation from biogenic mass is called biogenic DME ((Bio)DME).

We observe that the demand of zero emission leads to cycling and recycling of streams CO_2 and steam (H_2O) to (Reg)DME and oxidizing (Reg)DME to CO_2 and H_2O under using and applying energy and generating energy is a closed sustainable cycle.

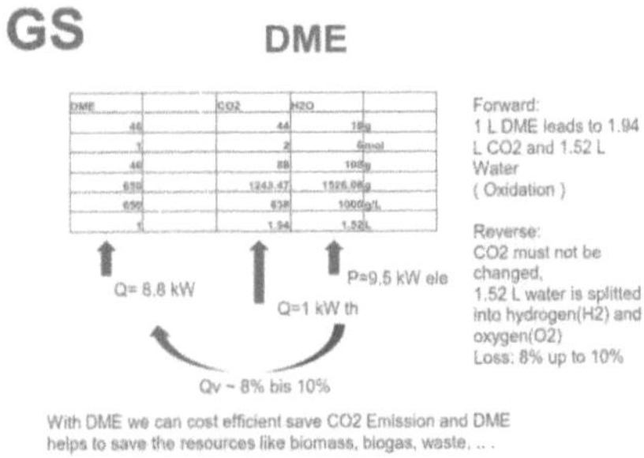

Figure 15: This drawing shows the energy costs recycling carbon dioxide and water with the production of dimethyl ether. (Source: Johann Gruber-Schmidt, 2016)

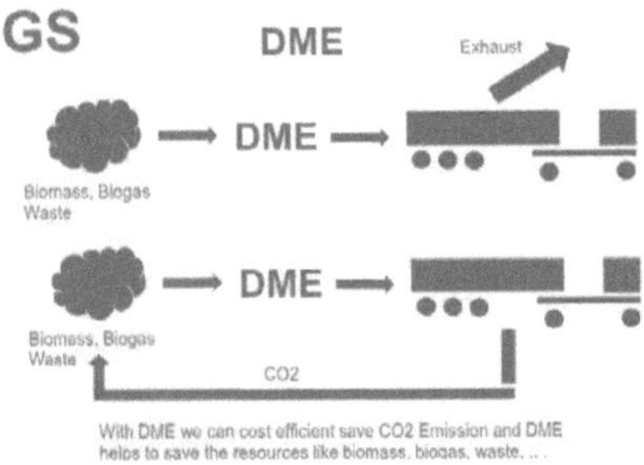

Figure 16: This simplest application of dimethyl ether to heavy trucks is substitution of fossil diesel with dimethyl ether. The costs for adaption of a bivalent fuel system are very low. The next adaption of the fuel system is saving carbon dioxide in a tank and recycling carbon dioxide and water for the production of dimethyl ether. (Source: Johann Gruber-Schmidt, 2017).

1.13 Fossil Fuel compared with dimethyl ether

The costs of DME generated from biogas with liquefaction over methanol synthesis and dehydration depend on the cost of the bioreactor biogas plant. The caloric low heat value (LHV) of liquid propane gas (LPG) is given with 12,9 kWh/kg ~ 6,6 kWh/L, the caloric low heat value of liquid butane 12,69 kWh/kg~ 7,6 kWh/L, depending on the gas mixture of propane, butane, nitrogen and carbon dioxide, the practical range of the low heat value of LPG 5,5kWh/L up to 6,6 kWh/L. The low heat value of DME is given in the range of 7,8 kWh/kg up to 8,2 kWh/kg, 5,6 kWh/K up to 5,8 kWh/L.

Comparing LPG with DME shows that LPG can be substituted by DME (Wobbe index: function of low caloric heat value and relation of density gas to density air):

$$W_o = \frac{H_u}{\sqrt{\frac{\rho_g}{\rho_{air}}}} \qquad [E\ 39]$$

	Propan	Butane	DME
Wobbe (kWh/L)	5,66	4,32	4,22
ρ_{air} (kg/m³)	1	1	1
ρ_G (kg/m³)	1,8077	2,4177	1,8977
H_u(kWh/L)	7,6	6,7	5,8

Table 3: Wobbe-index for propane, butane and dimethyl ether, comparing the Wobbe-index shows that the gases propane, butane and DME can be easily replaced by each other. (the technical effort is nearly the same).

Type	MZ	Kwh/kg
MeOH	32	5.50
DME	46	8.30
EtOH	46	7.40
DEE	74	10.83
Bt-OH	74	9.40
DBE	130	13.10
Biodiesel	198	9.50
Diesel	226	10.10

Alcohols		kWh/kg
MeOH	32	5.50
EtOH	46	7.40
Bt-OH	74	9.40
Biodiesel	198	9.50
Diesel	226	10.10

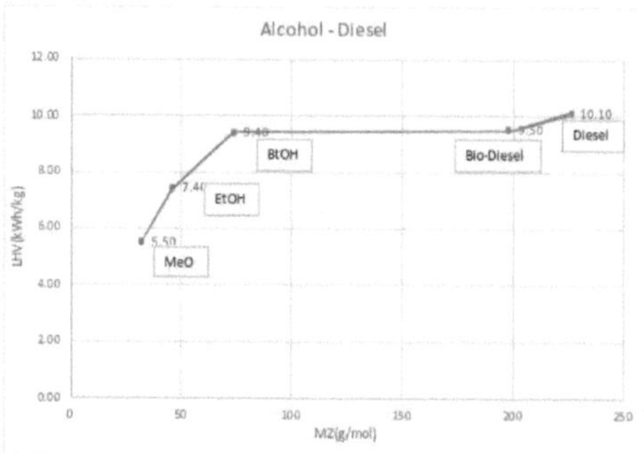

Figure 17: Comparing the caloric heat values of alcohols with bio diesel and fossil diesel. Data evaluated and converted from [23].

	MZ	Kwh/kg
MeOH	32	5.50
DME	46	8.30
EtOH	46	7.40
DEE	74	10.83
Bt-OH	74	9.40
DBE	130	13.10
Biodiesel	198	9.50
Diesel	226	10.10

Ether		KWh/kg
DME	46	8.30
DEE	74	10.83
DBE	130	13.10
Biodiesel	198	9.50
Diesel	226	10.10

GS

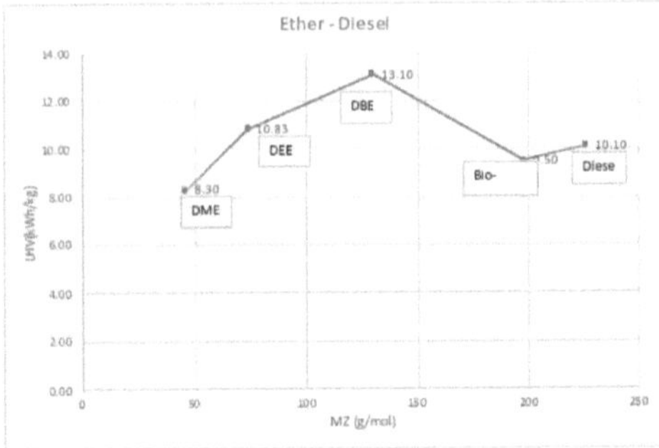

Figure 18: Comparing the caloric heat values of ethers with bio diesel and fossil diesel. Data are taken from [23].

The big difference: DME is a regenerative real Zero Emission fuel, LPG is a fossil product of the mineral oil processing. LPG technology like tanks, connections to flanges and tanks, pumps, valves can be applied directly to DME. Only the pressure controlling valves have to be applied to the different density and pressure.

1.14 Dimethyl Ether and Emission Trade System (ETS)

The emission trade system is an additional possibility to save costs. Investment in a adapted fuel system bivalent operating with DME and fossil Diesel, saves costs on the fuel tax, saves costs because of the lower price for DME and saves costs because of the possibility to sell emission certificates on the EEX bourse at Leipzig.

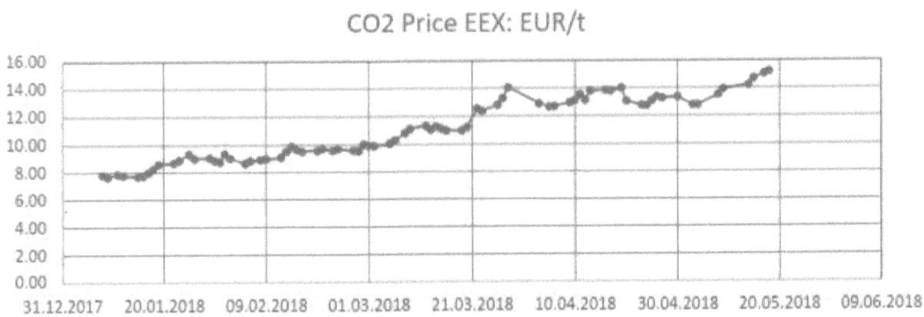

1 l Diesel ⇔ 3.09 kg CO_2
1 m³ Methane ⇔ 2.99 kg CO_2

Figure 19: The development of the CO_2 price on the EEX bourse at Leipzig (D) (Source: EEX Archive Bourse, 2018)

1.14.1 Application to a bus company

Bus: 30 seats								
						Diesel	DME	
1 Bus						Bh	Bh	
						2000	2000	
Type		P(max) kW	Q(kW)	Diesel	DME			
Mercedes		260	722.22	72.22	90.277			
		130	361.11	36.11	45.138	72222.22	90277.77	
		65	180.55	18.055	22.564			
Price				0.45	0.5			
Tax	20%			0.09	0.1			
Fuel Tax				0.45	0			
				0.99	0.6	71500	54166.66	17333.33
				EUR	EUR		24.242	
								Save Costs
35 Buses						2502500	1895833.33	606666.66
			Certificates					
ETS System		1 Fahrzeug						
1		Diesel	< = >	3.09	kg	CO2	15 EUR/t	
72222.2222		Diesel	< =>	223.166	t	CO2	3347.5	
								Gain ETS
2502500		Diesel	< = >	7732.725	t	CO2	115990.875	115990.875

1l Diesel = 3.09 kg CO2 1m³ Methan = 2.99 kg CO2

Table 4: This table shows saving costs using DME instead of fossil Diesel. Saving is ~ 30% of the operating costs. The investment for adapting the fuel system is lower 1% of the investment costs of a new bus. (Source: [2])

1.14.2 Application to transport company (heavy trucks)

In this passage we analyze the influence on the costs for transportation substituting fossil Diesel with dimethyl ether:

Cent/L

Diesel												
l/100km	Bh/d	l/d	d/w	l/w	w/a	l/a	Last	l/a.	Preis o. Ust	Preis o. Ust	EUR	EUR
45	10	450	5	2250	48	108000	100	108000	50	62	54000	66960
45	10	450	5	2250	48	108000	75	81000	50	62	40500	50220
45	10	450	5	2250	48	108000	50	54000	50	62	27000	33480

km/d	km/w	km/a
1000	5000	240000
750	3750	180000
500	2500	120000

DME												
l/100km	Bh/d	l/d	d/w	l/w	w/a	l/a	Last	l/a.	Preis o. Ust	Preis o. Ust	EUR	EUR
49.5	10	495	5	2475	48	118800	100	118800	45.45	56.36	54000	66960
49.5	10	495	5	2475	48	118800	75	89100	45.45	56.36	40500	50220
49.5	10	495	5	2475	48	118800	50	59400	45.45	56.36	27000	33480

dEUR	d EUR	d EUR	d EUR
-12960	0	0	12960
-9720	0	0	9720
-6480	0	0	6480
(Saving Costs Diesel - DME)			

External costs		(ASFINAG)					
Wien	Graz	km	EURO II	EURO V	100km	100km	d EUR
Wien	Graz	191	85.68	76.41	44.859	40.005	4.8534
Wien	Graz	km	EURO II	EURO IV	100km	100km	d EUR
Wien	Graz	191	85.68	76.41	44.859	40.005	4.8534
Wien	Graz	km	EURO II	EURO VI	100km	100km	d EUR
Wien	Graz	191	85.68	63.15	44.859	33.063	11.796
Wien	Graz	km	EURO IV	EURO VI	100km	100km	d EUR
Wien	Graz	191	76.41	63.15	40.005	33.063	6.9424

Savings, Emissionclass EU		
EURO III => EURO IV, V		
d	w	a
D EUR	D EUR	D EUR
4853.4	24267	1164817
3640.1	18200	873613
2426.7	12134	582408
EURO III => EURO VI		
d	w	a
D EUR	D EUR	D EUR
11796	58979	2830995
8846.9	44234	2123246
5897.9	29490	1415497
EURO IV,V => EURO VI		
d	w	a
D EUR	D EUR	D EUR
6942.4	34712	1666178
5206.8	26034	1249634
3471.2	17356	833089

GS

Table 5: This table shows the influence of using dimethyl ether instead of fossil diesel. The model is applied to a heavy truck running on the road under road pricing according to the demands in Austria. (Source: [2]) (At the end of the work is a bigger version of the table)

We watch that the benefit of using dimethyl ether instead of fossil Diesel is given by the following properties:

- Reduced mineral oil tax
- Reducing the external costs (road pricing)
- Additional gain from ETS (greenhouse gas emission reduction, certificates)

From this summary we learn, that it is not useful to look only on the heat caloric value and to compare dimethyl ether with fossil diesel or liquid propane gas. Doing this leads to a false conclusion.

The problems today are not only the exhaust emission and the pollution of the environment, this can be solved in an easy way to substitute Diesel with dimethyl ether.
The problems for transportation is the capacity of roads and the influence of the personal private traffic. This problem cannot be solved by changing the fuel, even if we change to electric source for the cars and trucks.

Using dimethyl ether can help reducing the emission of greenhouse gases, reducing the pollution of the environment. The benefits are the cheap costs of dimethyl ether compared to fossil diesel, the cheap application to existing and new diesel engines, the very high reduction of fine dust, exhaust gases like carbon monoxides and nitrogen oxides. If we look at the costs, application of dimethyl ether to heavy trucks is lower than 1% of the investment of a new truck, the reduction in costs for the read pricing in Austria, the same for middle Europe, is about 30%.

Additional consuming dimethyl ether in buses or heavy trucks enables companies to apply to the emission trade system and to save costs again. Applied to heavy trucks and to buses leads to a cost reduction about 30%.

Therefor dimethyl ether has a high potential to be used as an intermediate fuel in old and new diesel engines and in the development of robot drone applied to transport loads in the near-earth space.

1.15 Conclusion

Advanced Biogas plant prepared and structured as a bioreactor enables to optimize the biogas generation.

An alternative efficient possibility is the liquefaction of biogas to dimethyl ether (DME). There are two pathways: conversion over methanol and dehydration to dimethyl ether (DME), or direct conversion of the gas mixture carbon monoxide and hydrogen (1:1) to dimethyl ether (DME). Dimethyl ether (DME) can substitute fossil Diesel as a Zero Emission fuel, DME can also substitute as well fossil liquid propane gas (LPG) as a Zero Emission fuel.

1.16 Symbols and Short Cuts

The following symbols and short cuts are used in the text:

C_s	solid fixed carbon [kg/m^3]
C_{aq}	solved aqueous carbon [kg/m^3]
C_{xm}	methanogenic carbon [kg/m^3]
C_{co2}	carbon dioxide carbon [kg/m^3]
C_{xa}	acidogenic carbon [kg/m^3]
C_{ac}	acetate carbon [kg/m^3]
Y_a	acetogenic production rate [kg/kg]
μ_a	acetogenic growth rate [1/day]
Y_{ch4}	methane production rate [kg/kg]
K_{sa}	saturation constant for acetogens [kg/m^3]
K_{da}	death rate of acetogenic mass [1/day]
Y_m	methanogenic mass production [kg/kg]
K_{sm}	saturation constant methanogens [kg/m^3]
K_{dm}	death rate of methanogen mass [1/day]
μ_m	growing rates of methanogenic biomass [1/day]

1.17 Drawing and Pictures

Figure 1: Simple structure of a Biogas Plant operating on the basis of wet fermentation as shown in the project Hagenbrunn [2]

Figure 2: Combustion of biogas in well-known gas engines electric power range Pele = 500 kW_{ele} up to 2000 kW_{ele}, electric efficiency η_{ele} = 32% up 40%, biogas with methane concentration x CH4 = 40% up to 65%. (Source: [2])

Figure 3: The drawing shows the biogas cleaning route. (Source: [2])

Figure 4: Dry Reforming Reactor. In most cases dry reforming is realized in combination of the water gas shift reaction and several catalysts like Nickel, Calcium-oxide, Iron-oxide. (Source: [20])

Figure 5: Dry Reforming process reactions Equilibrium coefficient of the different reactions as a function of temperature. The temperature range is given between 500°C up to 850°C. (Source: [23])

Figure 6: Super Dry reforming with a combination of three catalysts, Nickel catalyst, Calcium oxide and Iron oxide enables to convert methane (CH4) and carbon dioxide (CO2) to carbon monoxide (CO) and steam (H2O). This concept can be enlarged and is strong controllable. (Source: [30])

Figure 7: Dry reforming in combination with chemical looping. Under chemical looping we understand oxidizing of a metal in the dry reforming reactor, deoxidizing in a separate reactor under heat supply. If the heat is generated by electricity we get a clean oxygen gas. If we generate heat from oxidizing methane we get carbon dioxide and steam. (Source: [30])

Figure 8: Drawing shows the steam gasification as realized in the project Traismauer [1] and Hagenbrunn [2], based on a top cycle (steam gasification) and a bottom cycle (gasification). (Source: [1])

Figure 9: equilibrium constants for the chemical reactions of steam reforming. The equations of [30] are calculated and plotted in a temperature range of T=500°C up to 900°C, and a pressure range p ~ 1.5 bar (nearly environment pressure). (Source: [35])

Figure 10: Low Pressure Methanol Synthesis as a function of pressure and temperature and the conversion of carbon monoxide (CO) and carbon dioxide (CO2) to methanol. The temperature range is given from T=190°C up to 350°C, pressure range from 50 bar up to 250 bar. In the biogas and biomass application we operate in the low-pressure range p=50 bar up to 100 bar to reduce the plant costs. (Source: [35])

Figure 11: The conversion of syngas to methanol in a reactor depends on the conversion rate (methanol yield). The heat of the product steam is recovered by the fresh input syngas. The offgas is gained after the condensation of the product steam mainly consisting of methanol and water. The offgas consist of hydrogen, carbon monoxide, carbon dioxide and light hydrocarbons. The offgas is recycled in the reactor again. (Source: [2])

Figure 12: Process of methanol synthesis – George Olah [30] applied to Traismauer [1] and Hagenbrunn [2]- this is a simple process flow, for the methanol reactor we have to apply the heat recovery and recycling of the offgas. (Source: [2])

Figure 13: Methanol synthesis as a function on pressure and temperature, pressure range p=10 bar up to 100 bar, temperature from T=130°C up to 330°C. (Source: [20])

Figure 14: Methanol synthesis as a function on pressure and temperature, pressure range p=10 bar up to 100 bar, temperature from T=130°C up to 330°C. (Source: [20])

Figure 15: This drawing shows the energy costs recycling carbon dioxide and water with the production of dimethyl ether. (Source: Johann Gruber-Schmidt, 2016)

Figure 16: This simplest application of dimethyl ether to heavy trucks is substitution of fossil diesel with dimethyl ether. The costs for adaption of a bivalent fuel system are very low. The next adaption of the fuel system is saving carbon dioxide in a tank and recycling carbon dioxide and water for the production of dimethyl ether. (Source: Johann Gruber-Schmidt, 2017).

Figure 17: Comparing the caloric heat values of alcohols with bio diesel and fossil diesel. Data evaluated and converted from [23].

Figure 18: Comparing the caloric heat values of ethers with bio diesel and fossil diesel. Data are taken from [23].

Figure 19: The development of the CO2 price on the EEX bourse at Leipzig (D) (Source: EEX Archive Bourse, 2018)

1.18 Table

Table 1: Table of measured and empirical values of the bio kinetic parameters for the generation of methane and acetogens. (Source: [24])

Table 2: Conversion of biogas to dimethyl ether. This table enables to get a better feeling on the relations of biogas, generated electric power, needed own electric power and produced DME (l/h). (Source: Johann Gruber-Schmidt, 2016)

Table 3: Wobbe-index for propane, butane and dimethyl ether, comparing the Wobbe-index shows that the gases propane, butane and DME can be easily replaced by each other. (the technical effort is nearly the same).

Table 4: This table shows saving costs using DME instead of fossil Diesel. Saving is ~ 30% of the operating costs. The investment for adapting the fuel system is lower 1% of the investment costs of a new bus. (Source: [2])

Table 5: This table shows the influence of using dimethyl ether instead of fossil diesel. The model is applied to a heavy truck running on the road under road pricing according to the demands in Austria. (Source: [2])

1.19 References

[1] J.Gruber-Schmidt, Project Traismauer, 2017, Presentation in Traismauer, Austria

[2] J.Gruber-Schmidt, Project Hagenbrunn, 2018, Presentation in Hagenbrunn, Austria

[3] K. Roel Westerp, 1989. *Neue Konvertersysteme für die Methanolsysteme*, Chem. Ing. Tech, 61, 3, 193ff.

[4] J.P.Syzbist, 2014. *Emissions and Performance, Benchmarking of Prototype DME Fueled Heavy Truck*, ORNL-TM -2014-59

[5] M.R. Rahimpour, et.all. 2011. *Reduction in CO emissions along a two-stage hydrogen-permselective membrane reactor in methanol synthesis process*, J. of. Industrial Engineering Chemistry, 17, 198ff.

[6] T.L. Roussiere, 2013. *Catalytic reforming of Methane in the presence of CO_2 and H_2O at high pressure,* Thesis, Nates, KIT

[7] M.C. Annesini, 2007. *Carbon Formation in Steam Reforming Process: A thermodynamic Analysis based on elemental composition,* Chemical Engineering Transactions,11, 21ff.

[8] Te. Chang, 1986. *Methanol Synthesis Reaction, Calculation of Equilibrium Conversions Using Equation of States*. Ind. Eng. Chem Process, 25, 277ff.

[9] Ahmed al Fatesh,2015. *Suppression of carbon formation in CH_4 and CO_2 Reforming,* J. o. King. Saud Univ. Eng. Sciences 27, 101ff.

[10] T.J. Siang, 2014. *Hydrogen production from CO2 Reforming of Methane of Cobalt based catalyst,* Thesis, Univ. Malaysia Pahang,

[11] R.Y. Chein, 2015. *Thermodynamic analysis of dry reforming of CH_4 with CO_2 at high Pressures,* J. o. Natural Gas Science and Engineering, 26, 617ff.

[12] Y. Zhang, 2015. *Catalysts for Methane Dry Reforming,* Advances in CO2 Capture and Sequestration, ch.8

[13] E.S. Van-Dal, 2013.*Design and simulation of a methanol production plant from CO_2 Hydrogenation,* J. o. cleaner Production, 57, 38ff.

[14] J. Schumann, 2015.*Cu, Zn-based catalysts for methanol synthesis,* Thesis, TU Berlin

[15] Chen, Xiachoun, 2000. *Dynamic Kinetics of Methanol Synthesis over a Commercial Copper based Catalyst,*Ch. J o. Chem. Eng. , 8, 315ff.

[16] K.M. van den Bussche, 1996.*A Steady-State Kinetic Model for Methanol Synthesis and the Water Gas Shift Reaction on a Commercial $Cu/ZnO/Al_2O_3$ Catalyst,* J. o. catalyst, 161, 1ff.

[17] M. Behrens, *Materials Chemistry of Cu/ZnO-based Catalysts for Methanol Synthesis and Steam Reforming,* Habilitation, 2013, TU Berlin

[18] F. Studt, et.al., 2015.*The mechanism of CO and CO_2 hydrogenation to methanol over Cu based catalysts,* J.o. ChemCatChem 7, 1105ff.

[19] E. Kunkes, M. Behrens, 2013.*Methanol Chemistry,* Chemical Energy Storage, De Gruyter Publication (R. Schlögl et. all.)

[20] J.A. Moulijn, 2013, *Handbook of chemical Process Technology,* J. Wiley & Sons P.

[21] Buelens L.C., Galvita V.V. et.al. 2016. *Super-dry reforming of methane intensifies CO_2 utilization via Le Chatelier's principle.* J. of Science, 10/12/2016

[22] VDI Standard 2048-1-2015. *Measurements, Uncertain Values and Measurements in energetic plants and power plants, basic theory*. 2015, BRD

[23] NIST, *Chemistry Book*, Database 69, 2018

[24] T. Amon, *Gasausbeute in landwirtschaftlichen Biogasanlagen*, KTBL, 2015

[25] U. Görisch, *Biogasanlagen*, Ulmer Publ., 2015

[26] J. Späth, *Analyse des Methanolmarktes*, 2015

[27] B. Höhlein, *Methanol als Energieträger*, FZ Jülich, 2008

[28] V.K. Vijay et. al., *Production of Methane through Anaerobic Digestation*, Abu Dabi, 2008

[29] P. Sabatier, *Heterogenous Catalysis*, Salzwasser Publ. 1914

[30] G. Olah, *Beyond oil and gas: the methanol economy*, Wiley Publ. 2006

1.19.1 Companies

[31] International DME Association, 2016. 1775 Eye Street, NW, Suite 1150, Washington, DC 20006

[32] VOLVO USA, DME Heavy Trucks, 2016, 13302 Pennsylvania Avenue Hagerstown, MD 21742, Ed. Saxman, Marketing

[33] FORD Motor Company, Customer Relationship Center, P.O. Box 6248, Dearborn, MI 48126, Infographic DME, 2016

[34] OBERON Fuels, 2445 5th Ave, Suite 200, San Diego, CA 92101, Jim Lane, A guide to Oberon Fuels, Simple Fuels, 2016, OBERON Fuels.

[35] J. Gruber-Schmidt, 2016. *DME a Rural Fuel with Zero Emission*, Klagenfurt, Austria, Agriculture Research.

[36] J. Gruber-Schmidt,2016. *DME a Zero Emission Fuel*, Budapest, Hungary.

Author
Johann Gruber-Schmidt

Table 5

Cent/L

Diesel												
l/100km	Bh/d	l/d	d/w	l/w	w/a	l/a	Last	l/a	Preis o. Ust	Preis o. Ust	EUR	EUR
45	10	450	5	2250	48	108000	100	108000	50	62	54000	66960
45	10	450	5	2250	48	108000	75	81000	50	62	40500	50220
45	10	450	5	2250	48	108000	50	54000	50	62	27000	33480

DME												
l/100km	Bh/d	l/d	d/w	l/w	w/a	l/a	Last	l/a	Preis o. Ust	Preis o. Ust	EUR	EUR
49.5	10	495	5	2475	48	118800	100	118800	45.45	56.36	54000	66960
49.5	10	495	5	2475	48	118800	75	89100	45.45	56.36	40500	50220
49.5	10	495	5	2475	48	118800	50	59400	45.45	56.36	27000	33480

km/d	km/w	km/a
1000	5000	240000
750	3750	180000
500	2500	120000

Savings, Emissionclass EU		
EURO III => EURO IV, V		
d	w	a
D EUR	D EUR	D EUR
4853.4	24267	1164817
3640.1	18200	873613
2426.7	12134	582408
EURO III => EURO VI		
d	w	a
D EUR	D EUR	D EUR
11796	58979	2830995
8846.9	44234	2123246
5897.9	29490	1415497
EURO IV,V => EURO VI		
d	w	a
D EUR	D EUR	D EUR
6942.4	34712	1666178
5206.8	26034	1249634
3471.2	17356	833089

External costs		(ASFINAG)					
Wien	Graz	km	EURO II	EURO V	100km	100km	d EUR
Wien	Graz	191	85.68	76.41	44.859	40.005	4.8534
Wien	Graz	km	EURO II	EURO IV	100km	100km	d EUR
Wien	Graz	191	85.68	76.41	44.859	40.005	4.8534
Wien	Graz	km	EURO II	EURO VI	100km	100km	d EUR
Wien	Graz	191	85.68	63.15	44.859	33.063	11.796
Wien	Graz	km	EURO IV	EURO VI	100km	100km	d EUR
Wien	Graz	191	76.41	63.15	40.005	33.063	6.9424

dEUR	d EUR	d EUR	d EUR
-12960	0	0	12960
-9720	0	0	9720
-6480	0	0	6480
(Saving Costs Diesel - DME)			